How to Become an Instrument Engineer

Part 1.523

New and Improved!
52.3% More Entertaining than Original

by
Gregory McMillan
and
Stanley Weiner, et al.

Illustrated by Ted Williams

Printed in the United States of America

INSTRUMENT SOCIETY OF AMERICA
67 Alexander Drive
P.O. Box 12277
Research Triangle Park
North Carolina 27709

Library of Congress Cataloging-in-Publication Data

McMillan, Gregory K., 1946-
How to become an instrument engineer part 1.523: new and improved! 52% more entertaining than original / Gregory McMillan, Stanley Weiner, et al. ; illustrated by Ted Williams.
p. cm.
Includes bibliographical references and index.
ISBN 1-55617-520-5
1. Engineering—Anecdotes. 2. Engineering instruments—Anecdotes. I. Weiner, Stanley, 1935- . II. Title.
620—dc20 94-16752
CIP

The et al.

Melisa Biri
Martin Gauthier
Bob Heider
Glen Mertz
Alan Neesley
Roger Reedy
Christian Toarmina
William Walvoord

Subtitles

"Stories by Instrument Engineers and Other Humorous People"

"Hot Tips on How to Survive Hot Times on the Job"

"How to Enjoy the Hero and Scapegoat Rollercoaster Ride"

"The Tao of Rocky and the Ti of Bullwinkle"

"Virtual Fun in a Virtual World"

Contents

Preface . . . vii
1 Rocky and Bullwinkle Get Real Tuning Loops . . . 1
2 Remembering Virtualman — of Mice and Men . . . 7
3 The Invasion of the Mind Snatchers . . . 11
4 Valve Position: the Missing Link . . . 17
5 Virtual Self-Test . . . 25
6 Frank Fame . . . 29
7 Reverse . . . 35
8 Top Ten Uses of Old Performance Reviews . . . 39
9 Top Ten Reasons to Suspect a Career Change . . . 43
10 Wayne and Garth, Vox in a Box . . . 47
11 How to Recognize a Bad Start-up . . . 55
12 Mass Appeal . . . 59
13 Liftoff! . . . 67
14 Dustin Arfman Retires from Stardom and Returns as An Instrument Engineer . . . 71
15 Combo 80 . . . 77
16 A Tribute to Stan the Man — The Existential Engineer . . . 81
17 The Interview . . . 87
18 Multiple Choice . . . 95
19 In the Good Old Days . . . 99
20 Topics from Dr. West . . . 105

Preface

The pen name of G. S. McWeiner, chosen to confuse our management, proved to be too effective. It confused everyone, including ISA. The first royalty check was held while ISA searched their files for a G. S. McWeiner. Also, ISA sections have been looking for a Mr. or Ms McWeiner for a guest speaker. We considered changing our names, but figured it was easier to use our real names for the next book. Finally, the preface in *Logical Thoughts*, developed at ISA's request to show that the value of the book just wasn't to have fun, set a somber tone that was not indicative of the attitude of the rest of the book. So, no more philosophy.

To share some fame and blame, we have chosen a few demented associates to contribute to this sequel. The results we believe are impressive. It shows that instrument engineers have a funny and creative side. We are considering a TV series or at least a rock video. If, after reading this book, you would like to contribute to the next edition, please send Greg or Stan your story. The royalties are used for a yearly party in New Orleans, or an other suitable site of superior process achievements for the contributors to discuss highly technical issues such as the composition of the best Hurricane.

1

TAI TANTALIZING TIDBITS
featuring MOOSE and SQUIRREL CUISINE
RESTAURANT
OPEN
...LOOK! THEY SPECIALIZE IN SERVING US!
HMM hum de DUM...
HUMAN RESOURCES GUY
YES, ON A PLATTER!

Rocky and Bullwinkle Get Real Tuning Loops

Gregory McMillan

The loops are oscillating, painting trend recordings with vivid colors and patterns. They belong in a museum as abstract art, especially since no one understands what they mean.

The engineers are busy designing plants for the Far East. They are sent to LA to help acclimate them to the air and noise pollution. Young engineers take this to mean there will be surfing overseas. Universities advertise: "Become an engineer and see the world."

In one company, it has obviously reached crisis proportions. The director knows about it. He summons his favorite manager and asks some disturbing questions like, "What the heck is going on?" In a brilliant flash of nonlinear thinking, he proposes a solution.

Director: Get Rocky and Bullwinkle to tune these loops.

Manager: You want me to get a flying squirrel and a moose?

Director: Do you have any people who can tune loops?

Manager: Well, they will work for peanuts and grazing rights, kind of like retirees.

Rocky and Bullwinkle are contracted but have a little trouble getting past the security guards at the airport.

Guard: What is with you guys?

Bullwinkle: We are going to Mardi Gras.

Guard: Not at this time of year.

Rocky: We are process control engineers.

Guard: That explains it. Go ahead, but stay out of the cockpit.

On board, after a cart load of bags of honey roasted nuts, Rocky and Bullwinkle read books on process control and get looped.

Rocky: This book on process control is the sorriest, dreariest, and dullest excuse for literature, I've ever seen.

Bullwinkle: You forgot "useless."

Bullwinkle: All of these books say loops are manually tuned by an open-loop graphical method.

Rocky: How many rulers have you seen in the control room? How many engineers do you see constructing tangents to inflection points? These books are just a propagation of a myth.

Bullwinkle: If you can't believe what you read, what can you believe? It's enough to destroy a young moose's faith. Here, I have written a tuning procedure.

Rocky: I didn't know you knew how to use WordPerfect.

Bullwinkle: Nothing to it.

Rocky: Zinflog lipstuk grassy sanwitch?

Bullwinkle: Of course, that's just a first draft.

Rocky: Here, put on our ultimate secret weapon, the Kirwood Derby used by Genghis Kahn, Alexander the Great, and Julius Caesar to conquer the world and Elvis Presley to conquer..., oh, never mind. It was last used by a Princeton professor who exclaimed

$E = mc^2$. It transforms anyone who wears it into the smartest person or moose.

Bullwinkle puts it on, configures the relay method for an auto tuner within the DCS and devises a manual closed-loop quarter amplitude oscillation method for difficult loops.

Rocky and Bullwinkle arrive in the control room and set the auto tuner to work on the loops with a lot of dead time. In some cases, the on-off control with the tuner is better than that previously exhibited by the loop. The tuner predicts higher gain, lower reset, and larger derivative settings. Where interaction or nonlinearity is a concern, the gain is cut in half. The noisy, sticky, fast inline loops are tuned by the manual procedure. For level loops, reset is completely removed. An error squared algorithm or gain scheduling is used for surge volumes where tight level control is undesirable. For batch to continuous transitions, an adapted velocity limit for a feedforward signal is configured in the DCS. The velocity limit spreads the step change in flow over the distance between the operating point and the maximum or minimum permissible level. The method is also used for surge tanks where multiple trains of equipment join to become the feed for a common train. For material balance control in column overhead receivers and residence time control in reactors, high gains (20 or more) are used with set point velocity limits and signal characterization for nonuniform cross-sectional areas (e.g., horizontal drums). Often, valves and actuators have to be repaired, because the process variable doesn't even change due to excessive valve dead band. Needless to say, we have one tired squirrel and one tired moose by the time all the loops are tuned and fixed.

Rocky: Let's be frank, we should do this for all the plants.

Bullwinkle: I would rather be Bullwinkle and let Frank do it.

The pair return to the headquarters and are notified that they are being transferred to Thailand. They refuse to go due to fears of appearing on a menu as the catch of the day. Instead, they make cartoons to teach process control.

2

ECSTASY...
OUTRAGE...
DELIGHT...
FRUSTRATION....
WARMTH...
LOVE...
HAPPINESS...
SADNESS...
KINDNESS...
CONTENTMENT....

Remembering Virtualman — of Mice and Men

Gregory McMillan

It has been about four weeks and it is still hard for me to believe Vandor Virtualman is gone. I was present at his retirement roast and brought the marshmallows.

Virtualman was constantly capitulating over his retirement plans and once told me, "The universe is a computer program without documentation." He reasoned that authentic being could be achieved on weekends, but it required a modem. He fretted that he would lose his log-in privilege and thus his identity.

I can still see him with his polyester gray slacks and white shirt. Preoccupied with weighty manners, he frequently forgot to let go of his mouse when he went to lunch. Fortunately, his computer was on a cart, so it could tag along. When I told him that a computer followed him to lunch, he said "Good, bytes in between bites."

One day, he showed up for work wearing mashed potatoes from trying to use a mouse on his wife and kids at dinner the night before. Coworkers perceived the potatoes as a possible major breakthrough in their personal relationship with Virtualman, and proceeded to pelt him with food at the lunch table. All agreed the food enhanced his wardrobe.

Virtualman lived in a virtual world. His constant banter about computers could be mistaken for coldness, but he was capable of great compassion. For example, after witnessing his boss get canned, he couldn't finish a second helping of doughnuts. Nontechnical communication was nonexistent. He preferred to have his most intimate conversations with a computer. Predictably, power captured his heart. For him, ecstasy was loading new software and finding the keys to the kingdom. (Most would be content with a castle. Some might not get past the moat.) His family had to use electronic mail to get his attention. His last and first communications each day, including holidays, were with the computer. As a legacy, he left the following tenets:

"Life is just a box of diskettes."

"My group consultant is a 286 with no output card."

"Who needs language when you have RAP (random acronym production)?"

3

...TAKE US TO YOUR INSTRUMENT ENGINEERS!
BIG ISA EXPOSITION
PIZZA & BEER
CHICAGO STYLE
BEER BEER BEER
PIZZA

The Invasion of the Mind Snatchers

Gregory McMillan

Flying saucers full of alien recruiters have landed and are luring instrument engineers on board with promises of advanced technology, no meetings, and coupons for deep dish pizzas.

Most important, the pizzas are not the cheap generic imitations where the crust tastes like recycled cardboard — we are talking about real Chicago style pizza with a filling that oozes cheese and delicacies from the stockyard — a feast for mind and body. The sight of one of these fresh steaming pies is enough to bring the strongest engineers to their knees. The alien managers are no dummies; they always first appear holding sample pizzas and wads of coupons. Their booths are the big attraction at the ISA show. Other exhibiters think they are from France.

In order for the aliens to travel to earth from another galaxy light years away and not run out of beer, they have to be a more advanced species, or very lucky.

The aliens' advanced technology eliminates devices that depend upon gravity. D/P cells have been extinct for centuries and prompt chuckles when mentioned by earthlings. Fiber optics, field effects, and the universal nature of waves are used for sensors and final elements. They are connected to photon parallel processing computers. Dead band from mechanical positioning is relegated to a museum of ancient horrors.

Most intriguing is the lack of alarms, buttons, or keys on the alien interfaces. Whatever decision that would be needed from a being is done faster with greater precision, logic, and knowledge by the parallel processors. Beams from special glasses allow the engineer to focus and select from windows in a three-dimensional space. The choices are the revisions to the configuration to continue to eliminate the need for alarms and manual actions.

The concept of a loop with its alarms on the process variable no longer exists. This is the hardest concept for the earth's engineers to grasp because on earth the advent of fieldbus with field controllers reinforced the paradigm of the loop. The aliens use first principles to unify the inputs and neural networks to adapt the outputs for the essential nonlinear interrelationships. Optimization of quality, capacity, and yield is simultaneous.

When asked about reorganizations and their paraphernalia, the aliens reply that there are no organizations. Assignments are done via fuzzy logic, chaos theory, and a dispatcher. The closest example on earth is the roving taxi cabs in a big city, where each driver individually, continually, and intuitively manages and communicates to maximize fare revenues by being in the right place at the right time. After all, how many goal documents, organization charts, and meetings do taxi cab drivers need?

When asked why they want lower-functioning earthling minds, the aliens reply that the potential of these minds remains largely untapped, and, to use an adage, a mind is a terrible thing to waste. The aliens view the earth as a toxic dump for intelligence, and they are into recycling.

Now, you might think the sudden disappearance of engineers and their families would cause a growing concern. Actually, the companies are delighted to lose their minds. They develop elaborate stories about transfers to calm the neighbors and associates. Each mind represents a cost savings of $100K per year. The stock price jumps a point for every hundred minds ditched. Eventually, the engineers are all gone. Wall Street is ecstatic.

4

YOUR COMPANY HAS JUST AGREED TO A QUANTITY DISCOUNT ON THESE STICKY-SLOPPY VALVES IF YOU OMIT THE POSITIONERS...
THESE ARE SO CHEAP THEY ARE WORTHLESS...

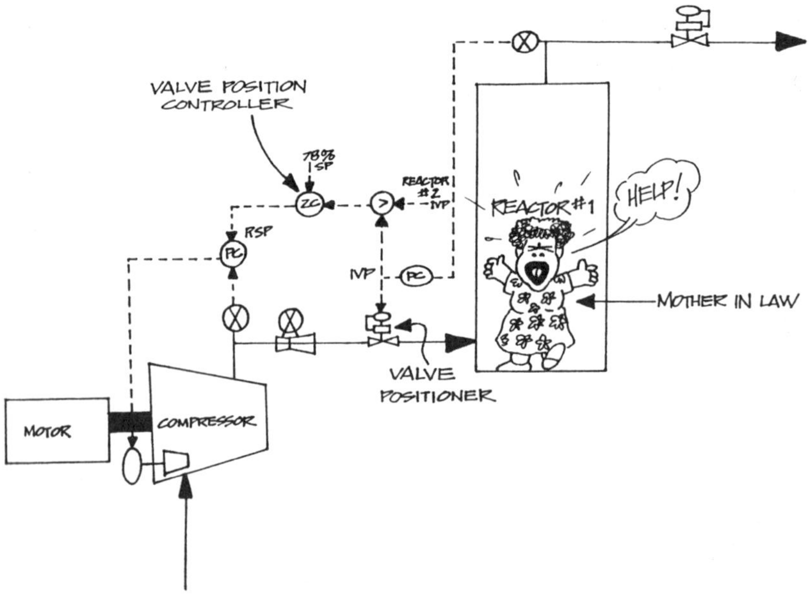

VALVE POSITION – THE MISSING LINK

Valve Position: the Missing Link

Gregory McMillan

Users are gradually becoming aware that control valves don't do what you expect, most control valves need to be serviced, the control valve is the only part of a control loop that directly affects the process, valve position is a leading indicator of the violation of an operating constraint or loss of control, and local optimization is possible by the addition of a simple valve position controller.

Position Perspective

Have you ever wondered what a missing link was doing at a crucial moment? For example, why did prehistoric man ascend from the cave to the condo, and was it the right move? Have you considered what your top management does at bonus time? The few times I have peeked into their offices, they have been doing the Charleston.

Are you conscious of what the control valve actually did during a severe upset? More importantly, how can I keep you conscious while you read about valve position? If your eyelids get heavy and you start to drift off into the land of tranquility, think about your mother-in-law, her moments of endearment, and whether she is just big or is hiding a valve positioner under her dress.

In the days of analog controllers and recorders, the controller output was seldom recorded or monitored except for a glance at a minimalist meter noted for sparsity of size and graduations. For a valve without a positioner, the meter readability was probably consistent with the stroke repeatability.

Imperfection Implication

Then, the distributed control system (DCS) came along with its ability to trend and use controller output, an *implied* valve position (IVP). Astute users trended IVP to detect valve size and performance problems, load shifts, fouling problems, and end points. For controllers with gains larger than one, pattern recognition for tuning was more noticeable in the plot of the IVP than in the measurement, particularly via a data historian connection at home after a few beers. But there was still the question of what the valve actually did. When you think about, the valve is the instrument that directly affects your process. The process measurement is an indirect inference of the result.

Dead band can be detected crudely by making small increments in the controller output and waiting longer than the process dead time to see if the measurement changes. Variable or unknown dead time, upsets, and noise make this difficult at best. The diagnosis of valve slip is even more testy, because the faster or larger changes in the measurement for an increment or decrement in controller output can be due to a decreasing time constant, increasing gain, disturbance, or interaction. Some orthodox religions believe suffering is the only way to redemption. One sect has been found to deliberately omit valve positioners.

The salvation — smart valves — has arrived but with a hitch. There is no fieldbus standard to make the valve diagnostics a normal part of the operator interface. Now, isn't this special? To get approval to spend the extra money, your project manager has to be locked up in a room for diversity training with the Village People singing "YMCA" until he/she signs diverse purchase orders. You guess which valves have to be smart and which could be dumb. Inevitably, you guess wrong, and you write change orders. The advantages of smart transmitters are becoming recognized. The smart positioner will prove to be even more valuable.

Some studies say fast loops should use boosters instead of valve positioners. The practical reality is that due to bench settings, nasty operating conditions, excessive booster dead band, positive feedback from high booster outlet port sensitivity, the high friction from environmental and graphite packing, and the large break away torque requirement of eccentric disc and ball valves, positioners are needed regardless of theoretical dynamic considerations. Also, smart positioners can insert a lag that eliminates the need to reduce the gain of a fast process loop.

Optimization Options

Let's pretend you have gotten past the issue of what the valve really did and you want to do some optimization, but you are not a multivariable sort of person. A simple solution is the valve position controller. This is not a positioner, which is a high gain proportional-derivative controller in a field device with linkages to the valve stem for position feedback. A valve position controller (ZIC) is an integral-only controller (no gain or rate) in your DCS whose measurement or process variable (PV) is a controller output. It is a standard loop point with a faceplate. The standard cascade functionality insures smooth, bumpless commissioning. The ZIC can be turned on or off by changing between the auto and manual modes. It slowly adjusts a set point to keep a valve at its minimum or maximum controllable position. Since the valve is an early indication of a load constraint before it affects the process, it can optimize without upsetting the process. For override control, PID with low gain action is used because integral action is stopped during integral tracking when the ZIC is not selected. Integral action multiplies the integral of the error by the *product of the gain and reset setting*. For integral-only control, the gain is one, and the contribution from the proportional and rate modes are suspended. Feedforward signals can be added to the ZIC output to help the normally slow ZIC to catch up with large disturbances.

(Wake up call! Your mother-in-law is at the door with suitcases. Quick, do you (a) feign death, (b) dismiss her as a traveling salesperson, or (c) act totally engrossed in this article and oblivious to external stimuli?)

An apparent contradiction in terms, the optimizations appear limitless. Some of the more obvious ones are listed below:

(1) Minimization of a compressor discharge pressure set point to force the furthest open feed valve to its maximum controllable position to save energy (see first cartoon).

(2) Maximization or minimization of a refrigeration unit's or a heater's outlet temperature set point, respectively, to force the furthest open user at its maximum controllable position to save energy.

(3) Maximization of a reactor's or column's feed set point to force the steam or coolant valve to its maximum controllable position to increase throughput when the constraint is heat transfer.

(4) Maximization of a reactor's air feed set point to force the feed valve to its maximum controllable position and take advantage of cool weather and nighttime temperatures to increase throughput when the constraint is air compressor capacity.

(5) Minimization of a kiln's stack temperature set point that determines secondary air flow to force the fuel valve to its maximum controllable position to increase throughput when the constraint is heat input.

(6) Minimization of the error in feedforward gain that needs feedback correction to reduce upsets when the controller output is summed with instead of multiplied by the feedforward signal.

(7) Minimization of an exchanger's temperature set point to force a cooling water valve far open enough to reduce fouling, process gain, and process dead time to improve the exchanger's response.

(8) Maximization of an incinerator's waste fuel set point to force an induced draft (ID) fan damper to its maximum position.

We could discuss these applications at length, but it appears my mother-in-law has just crawled in through the window and is positioning a valve over my head for optimal effect.

5

Virtual Self -Test

Gregory McMillan

This is a reality check. If your answer is yes to one of these questions, you should go on a vacation and not come back.

(1) At a magazine counter, do covers with pictures of personal computers instead of scantily clad people attract your attention?

(2) Do you use your computer while your spouse is in bed?

(3) When people ask "What's Happening," do you describe the status of your computer programs?

(4) Do you send electronic messages after 6:00 PM or before 6:00 AM?

(5) Do you leave for vacation with your computer and forget your family?

(6) Are your children named Picard, Gates, and Seagal?

(7) Do you keep time in terms of upgrades?

(8) Is the only sport you play or the only exercise you get on a screen?

(9) Are there more computer manuals than novels or biographies on your bookshelf at home?

(10) Do you look at others to know when to laugh?

(11) Is the only music you listen to computer generated?

(12) Is Amnesty International an organization to protect programmer rights?

6

"HI" GAIN
LES RESET
MOE RATE
SMOOTH FRANK FILTER
FAST FREDDY FEED FORWARD
PEPPY PETE POSITIONER
NASTY NORMAN NOISE
DASTARDLY DAN DEADTIME
HORRIBLE HAROLD HYSTERESIS

Frank Fame

Gregory McMillan

You can take great courses or read good books on process dynamics and control theory and still not have the slightest idea of how to get noticeable benefits from process control improvements. Frankly speaking, put the more esoteric stuff on hold. Changes as simple as the stick figures can lead to instant success. If you do the following, you might just be famous by Friday.

(1) **Decrease the reset action by a factor of 2 to 20 in all of your level, temperature, composition, and gas pressure controllers.** The more continuous gradual action of integral action lures humans into using too much reset. Reset is positive feedback that destablizes loops, particularly those with sticky control valves. The result is large amplitude excursions and overshoot and slow oscillations. For level loops and during set point changes to temperature loops, the reset should be less than 0.05 repeats per minute. New tuning rules such as Lambda by EnTech, drastically reduce the reset action in loops.

(2) **Increase the gain action by a factor of 2 to 5 in all of your level, temperature, composition, and gas pressure**

controllers on volumes after you have reduced the reset. The immediate action of the proportional mode will reduce loop variability. For surge tank levels, use an error squared algorithm or gain scheduling and a gain of two to four. For reactors, crystallizers, or distillate receivers where level control enforces the residence time or material balance, use a gain of five to ten. Larger residence times (large volume-to-throughput ratio) generally mean higher gains can be used. Thus, a typical temperature controller gain for a reactor is eight and for an exchanger is two. Similarly, a typical gas pressure controller gain for a reactor is six and for a column is two. Remember to just add enough of a measurement filter to keep the fluctuations of the controller output from noise within the dead band of the control valve.

(3) **Increase the rate action to at least 0.5 minute on all of your temperature loops.** The use of a thermowell will usually add a lag at least this large. Larger residence times generally mean higher rate settings can be used. For example, the temperature controller rate settings might be 2.0 and 0.5 for a reactor and exchanger, respectively. Two-phase level loops also need rate.

(4) **Use high gain positioners and throttle valves on all temperature, level, composition, and gas pressure loops.** These loops suffer the most from valve stick and slip action. Plug, eccentric disc, and ball valves are not genuine throttle valves. For best control, use piston or force–balance actuators. For liquid flow and pressure control, positioners create a slow inner or slave loop but are still needed for plug, butterfly, and ball valves, graphoil or environmental packing, steam service, and high pressure drops. The controller gain should 0.4 or less on liquid flow loops with positioners.

(5) **Ratio the flows of all loops that affect the same equipment.** The material balance on your process flow diagrams listed flows for various cases for good reasons. Enforce these relationships by simple ratio control where the set point of a flow loop is the filtered flow of another loop multiplied by a ratio factor. Don't forget the utility flows (e.g., ratio the wash water flow to the feed rate for a centrifuge).

(6) **Add feedforward action to all loops with severe but measurable load disturbances.** If the above is done and a loop still has excessive variability, track down the upset, measure and filter it, use a lead–lag and gain, and add it to the controller output. If you use a lead to compensate for a lag in the feedback loop's manipulated variable's path, make sure you use a lag time at least 1/8 of the lead time.

The above often involves just a change in a tuning parameter or the modification of a PLC or DCS configuration. The next suggestion takes much longer because it requires capital approval for additional control valves and transmitters. You might have to wait several months for fame and fortune.

(7) **Convert the manual and automatic on–off action of valves, switches, and pumps that upset important loops to regulatory loops with throttling action.** The most disruptive upsets are the fast ones caused by discrete actions. For example, sumps with level switches that turn pumps on and off should be replaced with an error squared level controller, and on–off weigh tank feeds should be replaced with mass flow loops. Regulatory loops also enable optimization by set point profiling, supervisory control, override control, and valve position controllers.

** The above are rules of thumb. Exceptions exist. Any change in tuning settings should be monitored closely.

7

"REVERSE TO GET THE INPUTS FROM THE OUTPUTS..."
NEURAL NETWORK
NEURAL NETWORK

Reverse

Gregory McMillan

Carnap the Magnificent removes envelopes from a mayonnaise jar, holds each one to his head and says, "Reverse and get the inputs from the outputs." The questions in the envelopes are as follows:

(1) Most performance reviews are a wasted effort because you can't ...

(2) Neural networks are relegated to roles as inferential sensors instead of control algorithms because they can't ...

(3) To compensate for a nonlinear installed valve characteristic, compute the required stroke for a desired flow to ...

(4) Controllers to get the flow to attain the set point ideally take the model of the process and ...

(5) Control theory assumes linearity and uses time constants in series to approximate dead time so you can get the inverse to ...

(6) To implement a classic full decoupler or dynamic matrix controller you take the inverse of the matrix to ...

(7) To get the feedforward gain you multiply the disturbance gain (i.e., change in percent of the controlled variable di-

vided by the change in the disturbance measurement) by the inverse of the open loop gain to ...

(8) To compensate a loop for a nonlinear process variable such as carbon monoxide, oxidation reduction potential, and pH, compute the desired flow ratio from the analytical measurement to ...

(9) Wall Street performance indices are counter productive because you can't ...

8

HOW MUCH ARE YOU WORTH ?
INTERESTS: STUDIES-10 IMPLEMENTATION-0
HAIR: CREWCUT-10 MOHAWK-0
EARS: PHONE HEADSET-10 STEREO HEADSET-0
TATOOS: COMPANY-10 SNAKE-0
HOBBIES: MEETINGS-10 SPORTS-0
APTITUDE: BUSINESS-10 TECHNICAL-0
BOOK: MEETING SCHEDULER-10 PROCESS CONTROL-0
TOES: FIVE-10 PAINTED-0

Top Ten Uses of Old Performance Reviews

Gregory McMillan

Now that we have officially found out that results reviews have no effect on either our raises or promotions (something we suspected all along) and that, from a lawyer's viewpoint, it would be better if they didn't exist, we need to creatively explore other uses for our files of MBO and MBR results reviews. Here are the top ten uses.

(10) Fire starters—for those fun times next to the fireplace or barbecue, reading manuals for the new simplified forms. Heck, you can use the manuals too!

(9) Insulation for your attic—the more paperwork you generated, the more energy you save. Go for a R50 factor.

(8) Christmas cards for your boss—staple the X ratings to your raise notices.

(7) Landscaping for gerbils—now *they* know how to have fun with paper!

(6) Gourmet delight for vegetarians—great fiber and there is an appropriate end.

(5) Puppy training papers—give a performance review on the puppy's ability to hit your performance review.

(4) Office decorations—highlight the really imaginative phrases.

(3) Targets for skeet—wad them up with glue and water and launch them.

(2) Science ficton—title it, "Engineers from Other Planets."

(1) Party hats for the next stockholders' meeting—sit in the front row and toss them on stage. Get your CEO to wear one.

9

WHAT DO YOU MISS MOST ABOUT YOUR JOB?
THE MEETINGS AND MY BOSS... I AM IN STUPIFICATION WITHDRAWAL...
THE DOUGHNUTS AND MY BOSS.... I AM IN LARD WITHDRAWAL...

Top Ten Reasons to Suspect a Career Change

Gregory McMillan

Since top management has gotten more creative in ways to eliminate people, we need to become more innovative in the recognition of impending alternatives to employment. Memorize the following:

(10) Office nameplates disappear into a hat.

(9) Your secretary asks, "What are you doing here?"

(8) A stranger is sitting at your desk.

(7) When you try to use your company credit card, it is listed as stolen.

(6) When you enter your log-on ID, the computer asks, "Are you joking?"

(5) The head janitor is seen in conference with your CEO.

(4) Plant management meets at an undisclosed location and never returns.

(3) It is getting dreadfully close to stock bonus time.

(2) Your CEO uses the word "sucks" ten times in describing your plant.

(1) You are asked to wear a price tag on your back when Japanese or Germans tour your plant.

10

VLASSIC PARK
OMIGOSH! THE GERKINS ARE LOOSE!
VLASSIC PARK
STAN NEVER TOLD ME HOW TO HANDLE THIS CONTROL PROBLEM!
...CRUNCH CRUNCH.....
BOY, AM I IN A PICKLE! OOCH! OUCH!

THE FUGITIVE ENGINEER
WOW!
KA-BLAM!
GET HIM!
IT WASN'T ME!
HEH HEH...
BILL'S LAPTOP

Wayne and Garth, Vox in a Box

Christian Toarmina

WAYNE'S WORLD!
WAYNE'S WORLD!

ENGINEER TIME!
ENGINEER TIME!

EXCELLENT!
EXCELLENT!

Wayne: Hello, I'm Wayne and this is Garth; and this is my cable access show, "Wayne's World."

Engineer on, Garth.

Garth: Engineer on, Wayne.

Wayne: O.K., I thought we'd start off today's show with a top five list.

Loud guitars chime in.

WAYNE'S WORLD!

TOP FIVE LIST!

PROCESS CONTROL MOVIES!

IS THAT ENOUGH FEEDBACK FOR YOU, DUDE?
WHOA, DUDE!
TED NUGENT Live
Disco IS DEAD

Wayne and Garth, Vox in a Box

Wayne: O.K., coming in at number 5 is TOP PROCESS. Tell the audience about it, Garth.

Garth: A cocky instrument engineer goes to work for the government on a nuclear aircraft carrier. He almost loses his edge when his best friend dies in a control mishap. Fortunately, he regains his composure, and his awesome control design saves the world for free enterprise and a capitalistic way of life.

Wayne: I think it's radical the way the engineers have code names in the movie. Vulture and Surf Man are really cool sounding names.

Numero quatro is ENGINEER GAMES.

Garth: A top rated engineer is forced by the CIA to take part in an international game of intelligence and espionage.

Hey, Wayne, what's your last name?

Wayne: Bond, Wayne Bond.

O.K., number three is THE FUGITIVE ENGINEER.

A process control engineer tries to prove his innocence in a chemical plant mishap. Dr. Erickson uses every control trick in the book to stay just ahead of the law.

Garth: It's hard to believe that a man with one arm could blow up a chemical facility, but with today's modern distributive control systems, if you can type on a computer, you can bring down a plant.

Wayne: Number two is THE CHEMICAL COMPANY.

Garth: A small chemical company in the south convinced an MIT senior to come to work for them with promises of a gold-plated pocket protector and a 486 DX/2 66 MHz computer. He soon discovers the company is run by the mob and there is no way out.

Wayne: No way.

Garth: Yes, way.

Wayne: Finally, my personal favorite, VLASSIC PARK.

An eccentric billionaire has a dream of building an amusement park of giant pickles and hires a brilliant control engineer to make his dream come true. All goes as planned until something goes wrong with the control program.

O.K., now is the time in the show where we take control questions from you, our most righteous audience. Please call us at 555-2353.

Wow, our first call.

Wayne picks up the phone.

Wayne: Hola, Wayne and Garth, control engineers extraordinaire.

Caller: Hello, my name is Greg McMillan, and I have a question about a cascade loop.

Garth: Are you the same G. K. McMillan who wrote *Tuning and Control Loop Performance* and *pH Control* for ISA?

Caller: One and the same.

Wayne and Garth immediately drop to their knees and start chanting: "We're not worthy, we're not worthy!"

Wayne: How can I be of service, oh exalted one?

Caller: We all know the basics of a cascaded loop. The output of the each controller is the set point for the next controller, except for the innermost loop. Control is improved because the inner fast loop quickly corrects for any disturbances, while, at the same time, the outer loop time constant filters the inner loop oscillations. I can go through the calculations and determine if a particular loop is a good candidate for an inner control loop, but can you guys come up with some simple rules in your own words so that any engineer knows when to use a cascade loop?

Wayne: Sure, here they are.

Guitars chime in once again.

Wayne and Garth, Vox in a Box

WAYNE'S WORLD!

RULES OF THUMB!

FOR EXCELLENT CASCADE CONTROL!

Wayne:

(1) Make sure the outer loop is a bigger dead head than the inner loop. (The outer loop dead time must be greater than the inner loop dead time. If it is not, increase the proportional band of the outside controller.)

(2) Make sure your guitars are tuned properly before a gig, otherwise you won't get paid. (Both the inner and outer loops must be tuned properly to realize any gains in control. Also, avoid integral action on the inner controller since it increases dead time.)

(3) Kick your background vocalist out of the band, if he is bogus. (Make sure the inner loop measurement has sufficient rangeability, accuracy, and speed.)

Garth: Wayne, I believe we're out of time.

Wayne: I comprehend, amigo. Until next time, engineer on, Garth.

Garth: Engineer on, Wayne.

11

...I GUESS THEY'RE A LITTLE BEHIND SCHEDULE...
HUBBA HUBBA!
COMING SOON... A NEW CHEMICAL PLANT
TODAY'S DATE... APRIL 1
STARTUP DATE... APRIL 30
MMMMMMooooooo....
BST SUPER COW
LAPTOP JR.

How to Recognize a Bad Start-up

Melissa Biri

Some projects are destined to have disastrous start-ups, and, no matter what one does, the path to destruction is impossible to avoid. But, for those of you who are new to this glorious problem, there are some telltale signs that, if recognized, can at least allow one to prepare for the inevitable **very long, all-expense paid vacation to a location not of your choice.**

You know it is going to be a bad start-up when:

(1) You are issued a frequent flier gold card the week before the start-up trips begin.

(2) You are hand delivering the contractor's copy of the demolition and construction drawings the day before the shutdown.

(3) Everyone on the project team but you schedules vacation to coincide with start-up coverage.

(4) The plant contact offers to find you an apartment in a nice location away from the plant but with two phone lines installed.

(5) The first day of start-up, you pull into the construction parking lot to find your very own reserved parking spot with your name prominently displayed by it.

(6) You are issued rain gear, heavy gloves, hard hat warmers, and long underwear, and it is only June in the midwest.

(7) All of the project team, except for yourself, takes early retirement six months before the project is scheduled for mechanical completion.

(8) You come back into the office during one of your first weekends off of start-up coverage and find someone else is assigned to your desk and computer.

(9) Your security badge to the home office no longer lets you in, but your plant security badge still allows you access to the plant.

(10) Your boss suggests that you buy your airline tickets in bulk or at least in groups of four.

(11) Your boss recommends that you get your picture taken so that your family can hang it on the wall at home, just in case they forget what you look like.

(12) The airline loses your luggage on the very first trip.

(13) Your flight never gets delayed going to the plant but always does when you are returning home for the weekend.

(14) Your boss recommends that you forward all your personal mail to the city you are working in.

(15) You are pulled over by a policeman within 20 feet of the airport exit because the rental car has expired license plates.

(16) Your construction engineer is arrested and fired during the checkout phase.

12

Mass Appeal

Robert Heider

A meter with mass appeal has a low pressure drop, great rangeability, can work on almost any type of fluid and linear output. This type of meter is in common use today, but not without some problems and watchouts. It is the Coriolis mass meter.

This meter measures flow through the measurement of forces due to the Coriolis effect. These forces react to a flow in a vibrating U-tube and act perpendicular to the flow. This force is proportional to the mass times the vector cross-product of the angular velocity of the tube and the velocity of the flow in the tube. An excellent article on the theory was published in March 1979 issue of *Mechanical Engineering*. This article will dazzle you with all kinds of calculus and vector analyses. After reading this you may wonder why you should use a meter you cannot explain to your boss. Just tell him that it is "the ultimate meter" both in price and accuracy and it won't plug. Talk to him about cross-products and he will probably just remember he is late for a meeting. A good way to observe the effect at home is to take a garden hose and dangle it with both hands in a three- foot U. Without any flow through the hose you will be able to swing the U in a oscillatory motion toward and away from your body. The hose will not twist. Have someone begin to run water in the hose while the hose is oscillating. The hose will twist, and you will not be able to prevent it. This twisting is the result of Coriolis forces. The amount of twist is the measurement made by mass meters.

This meter is accurate enough to replace weigh tanks and is probably the most expensive flowmeter on the market, but worth it compared to load cells. Just ask any structural engineer (if you can find any left). But with anything that good, where are the problems? For one thing, think about something vibrating like that in your plant—what about fatigue failure of the tube? Any mass meter company worth its salt should be able to tell you about this and how their design has been subjected to fatigue testing. This is one meter where you, the user, must be very specific about the process requirements. Chlorides are a real problem with stress cracking. The use of tantalum is cost- prohibitive, requiring board approval of most companies. Lined meters are now available for these problems. Another problem is two-phase flow. This is like bouncing a basketball half full of water. Another problem is fluids that amalgamate with the tube; this causes the tube to read the density incorrectly. Coatings and slurries are no problem, however.

Be careful of pulsating flow. It turns out that if the pulses are around the frequency of the tube's vibration, the meter will read erratic. This could be due to a piston or gear pump. This problem is rare but can occur.

Another feature of this meter is that you get density and temperature measurement thrown in for free.

Now that you have been sold on using one of these on that application that has had you stumped, you need to install it. These meters require some thought, and with careful attention to detail given to installation. Vibration is the most serious problem. Early attempts at overcoming this resulted in driving piles to bedrock, etc. Today, the designs make use of two U-tubes in parallel, one driven and the second stationary. This makes use of a balanced design and is less subject to vibration. But even though lab accuracy of this meter is great, you can shoot yourself in the foot if you don't follow instructions. Careful attention to detail should result in a good measurement.

Now for a history lesson. Just a few short years ago, there were two major players in this market—meter companies E and M. They both got into some heated marketing claims, worse than the beer wars. Both companies can trace their roots to two large St. Louis companies, and their CEOs were not on the best of terms, so to speak. This resulted in the good old all-American way of settling disputes—the court room. Lawsuits were filed in court and ended up at the Supreme. Meter E was owned by controls company F, which was owned by chemical company M. Meter M is owned by controls company E. Finally, chemical company M sold controls company F to controls company E and meter companies E and M were one. Controls company M did the obvious: closed meter company E's factory on the West Coast and consolidated operations at Boulder, Colorado. By the way, if you like to ski, meter company M has flow seminars at their Boulder headquarters. At your company's expense you can learn all about mass meters. Seminars in the Rockies are very common in the medical profession, so don't feel guilty about it when you go.

Per meter company M, the meter company E flowmeter was not as poor as meter company M claimed when they were in direct competition prior to the sale—surprise, surprise. (I don't think this is an issue, since both meters probably perform the same in the lab; installation issues should be the main concern.)

Examine the installation, paying particular attention to pipe supports and their distances on both sides of the meter. These supports should support the pipe on both sides of the meter, with the meter placed between the well-supported pipe. The objective here is to minimize the stress on the meter's pipe connections. Most pipe contractors don't like to be told how to support pipe, thank you. Herein lies the challenge. You must work with the tenacity of a bulldog to make sure the pipe is properly supported around the meter. This distance is critical to meter E's performance since, unlike the meter M, this distance affects the meter's resonance frequency. The supports should be solid in all three axes, i.e., the pipe should not rock or sway. Do not use threaded rods or similar-type pipe hangers; use angle stock or some other equivalent structural member. Stauff-brand pipe clamps should not be used to isolate the meter from vibration but as a method of ensuring good support to the pipe support. Meter E's manual stated that the pipe should be aligned within 0.03 inch. A good measure of the meter's installation stability can be determined by locking fluid in the meter and observing the density measurement, which should be stable within 0.0003 g/cc while the meter is pushed.

Another watchout is when the piping is installed: make sure that the meter is not used as a temporary support for the piping. This may cause the tube to be bent and not repairable. Make sure the tube is full of liquid so you won't end up with the basketball problem just mentioned. If you hydrostat the piping, make sure the meter is either out of the line or can take the pressure. We have had this problem, too. On meters that have to be traced and insulated, consider using a pre-engineered heat pack. This can save you a lot of work designing the right type of tracing and insulation design.

If you use two meters in series or several meters off a manifold, they may cross talk. Try to get at least 30 pipe diameters between meters. A quick way to check for this is to block fluid in the meters and look at the density signal on each meter. If there is a problem, this signal will be erratic. Crosstalk is when the vibrations from one meter affect the other. Getting around this requires "tuning" the meters, done with someone who has the skill mix of a tire balancer and a piano tuner. (Have you priced a piano tuner lately?) The idea is to place lead weights on the tube of one meter to adjust the meter to a different frequency. This should be done by someone from the factory.

Meter E is no longer made, and no tube repairs or replacements are available. Owners are given a discount for replacements with equivalent meter M. This is a good deal for meter E users to keep their installation up to date.

If all this fails, just box your meter up. No, not a cardboard shipping box to be sent back to the manufacturer but a steel box. This box should be a rectangular structure fabricated with 2" × 2" × 3/16" square steel tubing with four Stauff clamps, two mounted on each side to clamp the pipe. The objective of this support is to take up stresses around the pipe through the structure and not the pipe-to-meter connections. The sensor will then be supported entirely by the framework. Of course, the framework and sensor will have to be supported too.

Even in spite of these installation requirements, the mass meter's requirements are not nearly as difficult as those of orifice plates and vortex shedding meters. With a little thought and attention to details, this meter can be used with a minimum of 4 AM stays at the plant.

13

I'M A SPACE SHUTTLE!
NASA
WHOOT!
THERE IT IS!
LATER... BYE!
MY CAREER!
AHHWHOOP!
WHOOT! THERE IT IS!

Liftoff!

Bill Walvoord

All engineers, like most normal people, strive to achieve success in their chosen profession. Spectacular technical successes such as a space shuttle launch or a stroll on the moon come infrequently and are very rewarding to the engineers working on the project. Alas, we cannot all be NASA engineers and must try to achieve success in the sometimes not so spectacular world of instrumentation. However, with a little practice, ingenuity, and help from some of the other engineering professions, we can begin the journey to accomplish those spectacular successes.

It is wise to start out with smaller spectaculars and then work up. As an example of one of these small specs, picture the scene at a nondescript chemical facility with a bevy of engineers toiling away on a project. It is a retrofit installation in which chemical ABC was made for years and now chemical XYZ will be made using the same vessels. Of course, the project needs to be installed immediately. Since the process is not going to be run for an extended time period, everyone is striving to keep costs down.

A vessel is chosen as a holding tank for water in a scrubber system. It is a fiberglass tank with a flat bottom, a single nozzle at the bottom leading to the pump, and an assortment of nozzles on top. Level measurement is required. Since there is no spare nozzle at the bottom, the tank is open to the atmosphere, and a spare D/P transmitter is available, the choice is made to use a bubble system.

Assignments are made, and the instrument tech comes back noting that the level system is up and running. He/she is followed by the mechanical engineer who says that all the piping is in place, but what he/she does not say is that the tank is also totally blocked in.....even the atmospheric vent.

The countdown begins. Fiberglass vessels are not renown for their pressure rating...6...5...4...especially ones rated for no pressure...3...2...1... At about 2:00 AM technicians hear a loud AHHWHOOOPPP!! Liftoff!. Estimates are that the vessel, minus the bottom, attained an altitude of approximately 3 inches. Although it did not quite reach orbital velocity, this is a fine example of one of those smaller spectaculars that can lead to bigger things.

With a little more ingenuity, less communication, and more help from your fellow engineers, you could gain the benefits of having instant name recognition by the plant manager and the manufacturing director, having your name included in the monthly highlights and, with some luck, the summary of annual events, and having the opportunity to actually submit your resume to NASA (you've got proven experience, right?) as well as to several hundred other companies.

14

(...it makes perfect sense to us, and WE ought to know...)
ENGINEERING FLOW DIAGRAM
...HOW DID I GET OUTSIDE, FOR PETE'S SAKE!
WUPS! HI, DEAR!
COMPANY PICNIC
PLANT DRESSING ROOMS
PLANT EXCERSIZE ROOM... NOT RESPONSIBLE FOR ANYTHING
AARRGGH!
...They wouldn't laugh if they knew I was the major STOCK holder in this dump...
HEY! IF YOU SEE AN ENGINEER, GIVE HIM THIS! IT'S THE REAL ENGINEERING FLOW DIAGRAM!
WHOP!
EFD
EDDIE'S LAWN SERVICE

Dustin Arfman Retires from Stardom and Returns as An Instrument Engineer

Martin Gauthier

Dustin Arfman, distinguished actor, director, and producer has retired from the Hollywood scene after a long career in such movies as *Little Big Boy*, *Footsie*, and *Drizzleman*. Dustin is returning to his first love as an instrument engineer with what he remembered as a large, people-oriented, diversified, multi-sited, expansion-oriented, multinational chemical company headquartered in the midwest.

Asked by this reporter why he is attracted to the midwest, he said that the stubborn, hardheaded, single-minded, myopic vision needed for his *Drizzleman* character was developed when he previously worked for the same company. His previous association with process engineers and project managers was the perfect study guide for the character. The shuffling, cockeyed walk that he perfected in *Drizzleman* was actually learned at the firm through watching certain employees approach the manager's office for their annual results review session.

The *Little Big Boy* character, in particular, he said was easily developed while watching management continually outwit itself in decisions such as to develop a drink bottle and then to scrap it, to buy a large manufacturing site and then to sell it, to buy a large hardware company and then to sell it, and by spending megabucks on developing a product unwanted by the consumer.

The female impersonation for the *Footsie* character was, of course, one of his early efforts after a very long start-up at a southern plant in the early 70s. The many long days were balanced by many long nights in the French Quarter that left a lasting, but perverted, impression. The falsies that Footsie wore were an emulation of the Champagne girl, Rita Alexander, of French Quarter fame, who would finish her act by sipping from a glass of champagne perched precariously on one of her silicone masterpieces.

These trips to the French Quarter were necessary to relieve the extreme stress developed daily by unhappy but necessary associations with process engineers who swore that the production of the wrong molecule was an instrument and control problem. It was, of course, later discovered that a page of conversion steps had slipped out of the researcher's lab notebook. The researcher, it seems, had a late night attraction to an East St. Louis watering hole that more than equaled the French Quarter scene. This researcher, coincidentally, quit and appropriately went into comedy, eventually portraying a psychiatrist who developed a drug to return zombies back to reality. The failed molecule failed in the movie also.

Dustin, of course, has portrayed many unusual characters in his numerous films. He hope to do more character development while retraining as an instrument engineer. He expressed concern to this reporter about his lack of understanding of new hardware concepts. An integrated function controiler has distributed I/O!

The distribution control system uses integrated controllers! The field controllers, which he once so proudly moved from the field into the control room, are now being returned to the field. Most shocking of all, they're included as part of the valve! The property and protection people want to put sprinklers above motor control centers and distributed I/O cabinets! Dustin said that he feels that he's in a time warp and is considering a sci-fi film on his next vacation.

Although Dustin has not been back in this profession very long, he is considering the golden handshake that has been offered to him. However, there are so numerous character development opportunities at the firm, it's unlikely that he'll accept. He is particularly studying the new vice president who mumbles when he talks, is of short stature but tall character, and is known to give equally voracious chewing whenever a project is overrun or underrun! It's rumored that he also enjoys regular association with common folk over a good drink and is unusually perceived by lowly engineers to be a superior manager. His days are surely numbered.

But, back to Dustin. When asked why he left the instrument and controls field for the acting profession, his face showed his pain, and he related this tale: "I had spent a number of happy years in the central engineering group and decided I would try a stint in a plant environment. A transfer was arranged to a southern location with a promise of a return when requested. The plant world was certainly less structured than the corporate world. As an electrical and instrument project engineer, I expected to see engineering flow diagrams (EFD) or piping and instrumentation flow diagrams (P&ID) as part of the definition for the job. That rarely happened. Sometimes after start-up, a process or project engineer would produce one. By that time, I had bought and installed many unnecessary, or unwanted, or incorrect solutions and, surprisingly, sometimes a good solution.

"The engineers at the plants certainly respond very quickly to project requests, but documentation was considered superfluous by some. Design was done on the fly, by word of mouth, by point and shoot. My, boss, a respected Arkansas transplant, grew weary of these complaints from his electrical and instrument engineers. Finally, he decreed that no more projects would be executed by his group without first having an EFD in hand.

"That's where the problem started. As fate would be unkind to me yet again, the next project I drew was to be my undoing. It so happened that the board of directors for the corporation was scheduled to make a trip to this location for their spring meeting. A barbecue and reception was planned at the employees' picnic for the directors and their spouses. Well, we couldn't have these folks hiding behind the oak trees and in the bushes to relieve themselves or using the porta-potties as our female coworkers did. That became the overriding concern for the plant, and guess who was assigned to install the electrical and control portion for the new restrooms project at the plant picnic grounds.

"More importantly, the ultimatum for an EFD for all projects had just been promulgated. I stood my ground and out waited the mechanical engineer until the time was so close we almost didn't get the lights, fans, and pumps installed in time. But...I did get an EFD. It pretty much showed the whole process. I really was only interested in pump size, fan size, and lighting load. The EFD was a classic and would have been a good joke, except someone left a copy nailed to the wall in the women's restroom. The wife of the chairman of the board came out of that restroom waving it in the air as her husband was concluding his speech about the fine reception the board had been shown.

"So, if I can find a copy of the EFD, you decide whether I chose the right career path by quitting that day."

15

Combo 80

Christian Toarmina

Combo 80 for Monsanto is the magic number, when the combination of age and years of service totals 80, which enables one to get out of town with a nice stash. However, some of the other definitions that were suggested for Stan Weiner's early retirement are the following:

(1) A new ordering scheme at McDonald's: A hamburger and small fries for 80 cents.

(2) The combined total of your age and all the points you accumulated with the state highway patrol.

(3) A dance craze that peaked in popularity in 1980 consisting of a variety of "in-vogue" dance steps along with some limbo moves. Stan never could learn it, until now.

(4) Masterlock's latest invention: A safe with a new padlock of 80 digits to open it. Stan thought it was a good idea, so he bought one. Of course, he couldn't remember the combination, so Stan posted the numbers next to the lock.

(5) Comb O 80: The average number of strokes it takes a normal person to straighten one's hair. In Stan's case, it is comb O 3.

The following are the top six reasons Stan should be happy about reaching Combo 80 are:

(1) The ratio of Stan's retirement points to his age has gotten larger.

(2) Stan can fulfill his fantasy of playing the leading male in *Gone with the Wind* and, when asked about his results reviews, say "Frankly, I don't give a damn."

(3) No more orifice plates! No more orifice plates! No more orifice plates!

(4) Stan is now eligible for a free membership at Sam's Wholesale Club along with a senior citizen discount at K-Mart on Wednesdays.

(5) Stan can publish all the books he wants, including his newest: *How Combo 80 Influenced One Engineer's Life.*

(6) Stan can put strange little twists in the interlock guidelines, such as classifying an overflowing toilet as a class-4 interlock because of emissions.

16

A Tribute to Stan the Man—The Existential Engineer

Gregory McMillan

Stan experimented as a human data logger.

When asked to provide data logging, Stan rose to the occasion. He later realized he could sit back down if he bought an electronic one.

Stan pondered his first interface with management.

Stan kept on his desk a large supply of especially sharp looking RTDs. He was heard to lament, "So many targets, so little time."

Stan contributed to the success of bovine growth hormone.

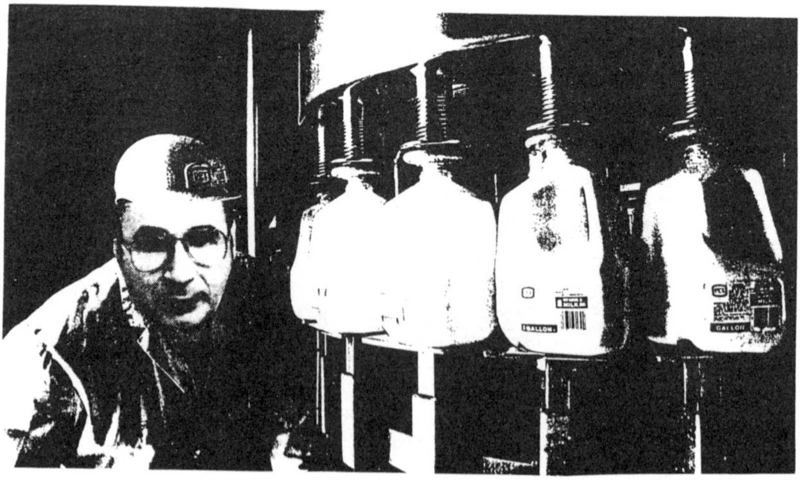

Stan was heard many times to tell his manager "Don't have a cow." Stan accidentally saw a cow when he got lost in Philadelphia and mistakenly crossed over into Jersey.

Stan is asked to go on a start-up that makes Desert Storm look like a picnic.

Stan attempts to program an all-terrestrial AGV to take his place, but it keeps going out for bagels and lox.

Stan experienced communication problems with DCS engineers.

He frequently yelled, "I don't care what size they are on the screen, the valves are too small."

Stan learned how to make high tech stuff work.

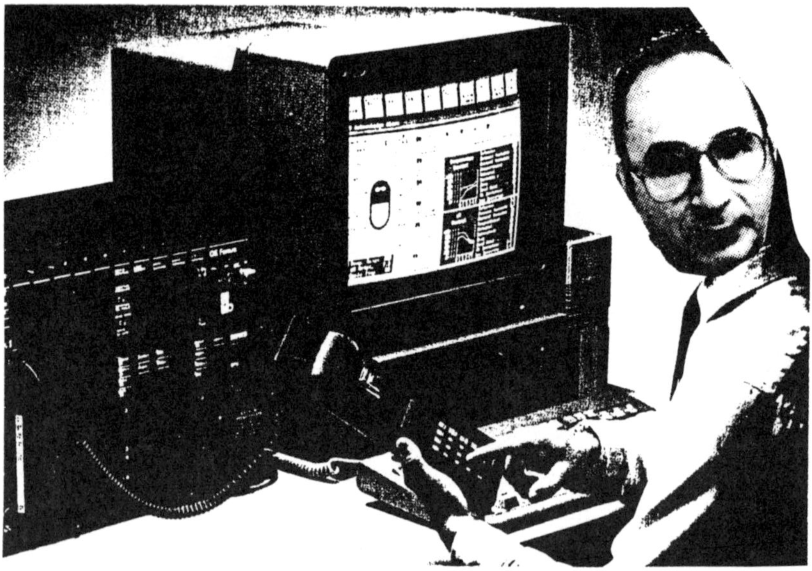

He spent a week configuring his TV and stereo and then tried out the same hand-held tuner on a DCS.

17

The Interview

Roger Reedy

(Extremely Highly Paid) Consultant: So, to conclude our interview, you've answered all of our on-a-scale-from-one-to-ten questions. Do you have any other ideas that may help your company attain 20% ROI?

(Control Systems) Engineer: Maybe. But I don't think I'd better tell you.

Consultant: Why not? Don't you want to help your company?

Engineer: Yes. But if I tell you, my career could be impacted. And it may end in a place where people talk about wind chill the whole year around.

Consultant: Your comments will be confidential. No one will ever know where they came from.

Engineer: Well, if you're sure. I mean, I'd like to get a load off my mind and help the company at the same time.

Consultant: Do it. Your company needs your help.

Engineer: Ok. I don't have all the answers, but I do know something about control. After all, process control is my business. And I do know this...if controls in my area worked like the financial controls in our company, I'd lose my reputation. Are you sure nobody will know who told you this?

Consultant: Relax. Nobody will know. Go on.

Engineer: OK. People in my profession have become more and more aware in recent years of a very important principle — that a smoothly controlled process is a good thing. Good control makes the company money. On the other hand, we pay a high penalty for trashy control. If a flow, for example, is allowed wild swings, that's a bad thing. This is because upsets at one point create disturbances that can result in process swings downstream. That can result in reduced efficiencies and reduced product quality. And that's not to mention the wear and tear on the control valve. A valve can be worn out in a matter of weeks if the control is wild enough.

Consultant: So if you recognize the problem, why don't you just go ahead and capitalize on it? Why don't you just fix the process problems and make more money for the company?

Engineer: We are, just as fast as we can! But I'm not talking about process control. I'm talking about corporate finance. It seems to me that the same principles apply. That is, a smoothly controlled financial strategy would be a good thing. Good financial control would make the company money. On the other hand, we pay a high penalty for trashy control. If a capital program, for example, is allowed wild swings, that's a bad thing. This is because it upsets people's workloads and messes up their schedules. Employee efficiency and the quality of their work will suffer if they are jacked around by rapidly changing finances. Consider the poor engineer who is blown about by every change in the wind. Why, a person could wear himself out trying to adjust for all the financial disturbances.

Consultant: What do you mean, "disturbances"?

Engineer: OK. So take, for example, the way a recent project was handled. This project was approved for a guest operation in

October. Due to management by cash flow, the equipment had to be delivered and the money had to be spent before the end of the year. This was not because the equipment was needed, you understand, but because the money had to be *spent by a certain time*. We didn't need the equipment until the following May. Anyway, I got a commitment from the Instrument Company rep and he got the commitment from the Instrument Company to meet delivery and to meet our requirement for invoicing. This was like a world record delivery! Imagine how silly I felt when, immediately after confirmation, I had to call and tell the Instrument Company rep we had a change in our requirements. Another plant had spent more money in the year than scheduled and now we couldn't spend any until next year! He very graciously agreed to hold the invoices and not make us overdue so we could pay after the first of the year. You see how inefficient that was for us and for the rep and for the Instrument Company? I created inefficiencies for myself and others by pushing the authorization through. I rushed and jerked the Fisher Rep around, getting the quotations out too quickly. I rushed to get the purchase orders approved and pushed Purchasing to get them placed. I almost wore myself out doing things and then undoing them in the rush. Guess what. Before the end of the year, things changed again. I had to go back and ask for the invoices just in case we "needed" to spend some of the money before the end of the year. On December 31, we paid 40%. Whew! I'm worn out just telling the story!

Consultant: I think I understand. What do you propose to do to make things better?

Engineer: Not yet! Let me vent more pent-up frustration first. This feels like good therapy.

The capital budget is a very interesting control situation. Most years, money is tight at first. Then, sometimes late in the year, the lid is opened and managers come asking where can we spend money. Could that kind of "control" be efficient with quality results? No! If we want to make a 20% return, we have to be more consistent in how we do our work. We need consistency so we can even out our work load and set schedules that make sense. The quality of our work depends a lot on how well available funds are administered.

So what can be done, you ask.

Consultant: We need to know that.

Engineer: First, money managers need to recognize what control system engineers have discovered — that a well controlled process will yield unexpectedly high returns. *Recognizing* that fact is the first step. Then, I guess some modeling is in order. This is a gross oversimplification. Available funds in a program is comparable to level in a tank. New funds is flow in; money spent is flow out of the tank. Accurate measurements of flow in, flow out, and level are necessary to make the controls work. There may be some external influences that need to be measured and integrated into the strategy using feedforward, perhaps. Filtering of the measurements may be needed to smooth out effects of the discrete nature of money flow. And after the model is complete, good controller tuning is a must.

For a tank with flows in and out, proportional-integral-derivative control works well. However, for the financial model, PID may not work at all. There may be some algorithm already invented that will apply. On the other hand, there may not be.

Consultant: So, what if there's not? How is what you've told me supposed to help?

Engineer: If for whatever reason the process control system on our real-world tank does fail to control on automatic, the Operator will put the renegade loops in manual. In this mode, he/she must carefully watch the measurements and manually change the position of the valves to maintain the level and the flows within reasonable limits. Some operators are naturally better at running a process in this mode than others. But in order to do a good job, he/she must know that it's important to make the operation smooth and must have the skills to react in the right ways.

In the absence of an automatic financial controller, the "Financial Operator" is continually making adjustments in the manual mode-based available inputs. It would help a lot if these "operators" recognize that smooth financial control is a good thing.

Consultant: Makes sense to me. Your comments will be included as part or our report. And don't worry...the source of this information will be strictly confidential.

Engineer: That's OK. I've changed my mind. You can tell who it came from. Saying all that felt right. In the immortal words of some great person in history, "I'll fall on my sword for the good of my company.

Consultant: Well, OK then. I guess I've kept you from your work long enough. You probably need to be out making some controls work better.

Engineer: I'd like to do that, but I don't have time. I have to rewrite a scope report to convince management to approve money for a cost improvement project. We're into the "hard" season, you know.

18

A. SHMUCK,
ENGINEERING
DUDE
DATA BASE
CORRUPTED
HAVE A NICE DAY...
EASY
SOFTWARE
UPGRADE

Multiple Choice

Martin Gauthier

You've just updated the flash ROM in your DCS controller with the new operating system that fixes these bugs and provides those enhanced features you've yearned for. Now the controller won't work, and the manufacturer says you need to perform a total download to your batch controller.

Pick the most appropriate response from this list:

(1) Use very colorful creative language as you hang up the phone.

(2) Threaten to buy a different DCS at the next opportunity.

(3) Get an empty feeling in the pit of your stomach as you approach production management.

(4) Sweat profusely, tell the boss you're sick, and that he/she should call your worst enemy to fill in.

(5) Call your spouse and cancel all plans for the weekend.

(6) Call all your DCS friends and invite them to a party.

(7) Find that the database is corrupted.

(8) Your database computer decides to take an early break for the weekend and the vendor says he can't come until next week.

(9) Decide that a career change should come now.

(10) Submit early retirement papers.

(11) Wish you were a process engineer.

(12) Wish you were a sanitary engineer.

(13) Wish you had stayed at that summer job as a manure spreader.

(14) Your boss squared appears and asks how it's coming.

(15) All of the above.

19

DAILY PLANNA
RIP VAN WEINER
AN ACTION BOOK
MAALOX
ECONO-SIZE

In the Good Old Days

Stanley Weiner

I really don't think of myself as physically or mentally "old."

I play one and a half hours of racquetball three or four times a week with younger and bigger people. I ride my mountain bike when the temperature is above 55 degrees. I have this fancy color lap-top computer that has cute little pictures and can tell all about the side effects of drugs and prints my legal documents. My car has a 230-HP flat six engine that will do 149 miles per hour (the book says), all- heel-drive, an exotic stereo system, windows that open in the rain and you won't get wet, other windows that won't open even though they look like they should, an air-conditioning system that measures the outside temperature and the sun intensity, and on and on. The house has two stereo systems — one normal. The other has six-foot high speakers that look like they came from the movie *2001*. The two amplifiers are rated at 850 watts each. There is a digital cable system that plays 33 different kinds of music with no talking. The number of stereo VCRs has been reduced from 5 to 2. Maybe I am old! I didn't have all this stuff when I was young. But anyway, IN THE GOOD OLD DAYS.....

At the end of every month, we took a piece of paper called a "time sheet," wrote some code numbers that identified what projects we worked on, put some hours in for each code—making sure it added up to 168—and gave it to the secretary. Every once in a while, a manager sent you a nasty note about charging too much time to a certain project. You could complete the sheet in about 3.59 minutes while drinking coffee. Then someone in INFORMATION SYSTEMS decided to AUTOMATE THE SYSTEM. After automation, it takes 3.59 minutes for the computer to scan for viruses, 38 seconds to check your password, 2.6 minutes to get the correct menu, 2.4 minutes to connect to the LAN, 1.3 minutes to get the LAN menu, 13.7 minutes to fill out the time sheet (provided you don't screw up the code numbers), and after all that works, you need to get out of the system. On a good day, one to two hours could be wasted on "filling out a time sheet.

Expense accounts were always easy things for engineers to fill out. To make plane, hotel, and car reservations, you called the designated travel department or agency. They took care of all the details and sent you the tickets, boarding passes, confirmation numbers, parking vouchers, and corresponding receipts. They charged everything to some magic credit card that you never saw. The secretary got you some cash. While on the road, additional cash would be available from any company location. When you returned, you filled out an expense account form and settled up on your out-of-pocket spending. The system was simple, you didn't have to use your money, and it worked.

Then they (the accountants) discovered computers. Everyone gets their own charge card. No more cash advances. If you want money, you get a limited amount from a machine at the airport. When you return from a trip, you turn on the computer and, if you enter the correct code number in the proper sequence, go through a procedure that is similar to filling out a time sheet. If the bills come in before the expense account check was received, you get nasty letters from the charge card people saying that you are irresponsible and lazy and are building a poor credit record. In order to avoid some of these problems, people turn in partial expense accounts to cover things like airplane tickets. Maybe the paper system wasn't so bad.

Most of the time, specifying and buying instruments was relatively easy. If you had some questions, you called up an instrument sales person who brought you specification sheets, prices, and delivery information. Most of them even knew something about what they were selling. There were even occasions when they took verbal orders and/or a hand-written specifications and sent the instruments or valves to the plant in a cab or delivery truck. Along came Mr./Ms Computer.

Now, instead of getting a blank specification form from a filing cabinet, you turn on your trusty computer terminal. Again, check for viruses, get a menu, enter the LAN system, get another menu, enter codes to get into the specifying and purchasing system, and then hope that someone else isn't in the same project area modifying some existing work. On and on and on..............

Has anyone ever tried writing some basic information on the back of a management memo, giving it to a vendor, and letting them generate all the paper that is required to specify, purchase, build, ship, install, and maintain the equipment? Or if you really like computers, you could put some basic information in a word processing program, give the vendor a disc, a completed disc would be returned, and the equipment delivered.

IN THE GOOD OLD DAYS, engineers would sit at their desks with a pencil, drawing pictures and filling out forms. Once in a while, they would take a walk over to the drafting department to see what was going on. Now the obvious difference between an engineer and a drafts-person is the size of their CRT and keyboard (engineers get smaller tubes). Now, everyone gets a computer. Everyone sits at their desk staring at the colored lights, or they go to meetings to discuss better ways to stare at the colored lights. There was a time in the 1960s and 70s when people who stared at colored lights were suspected of being on drugs; however, we know that's not the case today. Everyone knows that engineers don't take illicit drugs, and, besides that, they are periodically tested to find out that they don't take drugs. Some companies continue to spend millions of dollars looking for "engineering drug addicts" even though they have never found one. When middle-level management is asked about the drug testing program, they respond that it is a goal of upper management to have a "drug-free engineering department."

20

GEE... I NEVER THOUGHT ENVIRONMENTAL WORK WOULD INCLUDE ED's COFFEE!

Topics from Dr. West

Stanley Weiner

(1) The Helpee-Selfee Chapter for Ph.D.s. and Other Flakes

Let me start out by saying that I would very much like to have a Ph.D. Fortunately or unfortunately, however you want to look at it, at 23 years of age, I got tired of being poor and decided six years in college was enough. So, I went to work where I had contacts, in different capacities, with a variety of people with Ph.D.s. I can't really recall that these contacts and/or relationships were anything that concerned me or caused me to think much about them.

About the age of 40, something strange happened. Someone whom I had known as a child on the East Coast moved to the Midwest to begin doing medical research at a large university. He had a Ph.D. and had done post-Docs in the United States and Europe. We spent hours discussing economics, politics, science, music, motorcycles, why the rich people don't want to give money to the poor, and other related topics. These discussions led to an association with a college professor who introduced me to oriental philosophy, meditation, and art. Next I became friends with a "mental health professional" who had a Ph.D. in psychology. We spent a great deal of time talking about scientific theories, the meaning of life, why trees know so much, communication, how people get themselves screwed up, and other related topics. The stories go on and on. Currently my significant other, best friend, old friends, daughter, son-in-law, and many of their friends have Ph.D.s. This means that many of the social events I attend have a very large percentage of Ph.D.s. Try having lunch with three Ph.D.s. They are trying to decide if all engineers are crazy, strange, stupid, or together they are just trying to make them look crazy, strange, or stupid.

Anyway, out of all this comes the following Helpee-Selfee Rules:

- If you want to be a real Ph.D., be prepared to spend many years.
- You must believe that almost all the answers are in some book or technical journal.
- If the whole answer is not in a book or technical journal, someone started to think about it but didn't finish because they either ran out of time and/or money.
- If someone started to think about it and ran out of time and/or money, they at least wrote some of it down in some book or technical journal.
- Don't ever start to think about something until you check with a good library that has books and technical journals.
- Book stores are wonderful places to spend your time—buy books and magazines.
- Fleamarkets, used book sales, and garage sales are wonderful places to buy books.
- There is no relationship between the number of books that you buy and whether you read them or not. Just having them is the important thing.
- It's OK to ask questions provided you check out the answer in some book.
- Engineers with Ph.D.s are not real engineers.

Topics from Dr. West

- Don't believe what engineers tell you—they make up stuff that can't be found in any book.
- You must have a Ph.D. to be a real scientist. People with undergraduate degrees in chemistry, physics, biology, mathematics, etc., are lab technicians.
- You must have a Ph.D. or M.D. to help people with mental problems. If you have a master's degree, you are a "social worker."
- Ph.D.s. from business schools really don't think much of M.B.A.s.
- M.D.s. don't like Ph.D.s. doing medical research.
- There was a Ph.D. who had 13 VCRs in his house—12 were stereo and one was mono for his wife to record soap operas.
- Ph.D.s. like hobbies—too numerous to list. However, if the hobby becomes something that needs to be done regularly, they ignore it until they forget that it has to be done.
- Ph.D.s. are very opinionated. They don't like most things.
- Ph.D.s. do things that they like to do first. They don't prioritize things like engineers do.
- Ph.D.s. like to use words and phrases that other people don't understand. Sometimes they say things that even other Ph.D.s don't understand. Ph.D.s. have a problem with a nonunderstanding person, including other Ph.D.s.
- Completely explaining something to a Ph.D., explaining how important it is, and requesting that they do it means nothing.
- Ph.D.s. do not work regular hours—like 8 AM to 4:30 PM.
- Time is a variable that is variable depending on the circumstances. "I am leaving in 5 minutes" does not mean that in 5 minutes, a Ph.D. will put on his/her coat and move towards the door.

minutes, a Ph.D. will put on his/her coat and move towards the door.

The following is a true Ph.D. story that illustrates some of the Helpee-Selfee Rules.

After the "really big war" was over, a Ph.D. physicist was made the director of a research and applications laboratory for a large company. The lab had a large, diversified staff and many functions. They had an electronics development group that worked on computers and data acquisition systems. There was an analyzer development group. However, the major effort was devoted to instrument development, testing commercially available hardware, and solving problems that the plant engineers did not understand.

For simplicity, let's call the lab head "Big Ed."

Big Ed liked to drink coffee. The first problem was to find his cup. Since the lab was large and in sections, Big Ed would wonder around, talking to the scientists and engineers about their projects; however, he would get so wrapped up in the discussions, he would walk away without his cup. When he realized that he needed some coffee, the cup was difficult to locate. After about 6 months of listening to his ranting and raving about his lost cup, a group of us recommended that he put 12 empty, clean cups in his office. Initially, when he wanted a cup of coffee, he could take a new, clean one from his office, go to the coffee pot, fill it up and start his rounds. Later in the day when he needed another cup, he could get another clean cup from his office. After a few days, there would be cups spread all over the lab and he wouldn't have to walk back to his office. After listening to our idea, Ed went to the library to find out if this had ever been tried before. He spoke to Ph.D.s. in other disciplines, such as chemistry, and decided that it was at least worth a trial period.

The only really disgusting part of this idea (which we didn't explain) was the fact that Big Ed never washed out a cup. He just continually added fresh coffee to the cold stuff that was in the cup. It was obvious that he had not looked in any biology books or he would have discovered that things grow in warm organic liquids.

At the beginning of the experiment, things went reasonably well; however, unanticipated problems began to occur. Even with multiple cups, there were times when none could be found. When one with green stuff growing in it was left on someone's desk, we suspect that it was thrown in the trash.

And then a really serious problem began. Ed started to come to work close to the normal starting time. He would walk around cursing until he found a cup. Then he would go to the 40-cup coffee pot and start to fill the cup. When it was full, he would look at the contents and say something like, "Coffee looks kind of weak this morning." Someone standing by the pot would say, "The coffee isn't ready yet—you can tell because the little red light hasn't come on." This would thoroughly confuse Big Ed. We tried getting electronics experts, instrument engineers, technicians, and the secretary to explain that the red light on the side of the pot means that the coffee brewing process is complete and you could fill your cup. Unfortunately, none of us had the communication skills to get the message across.

In desperation, we designed a timer system to fill the pot with water and start the brewing process an hour before the normal starting time. The coffee was ready when Ed arrived, but he still couldn't find his cup. To make things worse, he started bringing apples from home, taking a bite from one and leaving it around the lab. Two or three hours later, be began to look for his partially eaten, rotting apple. He was convinced that a deranged instrument engineer was out to get him.

I eventually transferred out of research into the engineering department, where I would like to say that I lived happily ever after.

(2) When Was the Last Time You Tried Changing the TRANSMISSION RETRY VALUE IN THE PRINTER'S CONNECT DIALOG BOX?

IN THE OLD DAYS, before the revolution, we bought computers for about $450,000 each. They had 48K of RAM and some kind of (you should excuse the expression) mass memory system that was slow and/or forgot what it was supposed to remember. We also had "computer experts" who talked to these beasts in something called "machine or assembly language." One good part of this story was that normal engineers had nothing to do with the machines and paid very little attention to how they did their "thing." Another bad thing was that you needed a room 20 feet by 20 feet to hold all the equipment and the manuals for the programmers. The heating and ventilating (HVAC) engineers sized the air conditioners by calling IBM or GE and asking them, "How much heat do the machines generate?" They usually got a number that sounded something like 30,000 Btu/hour. They figured that the outside temperature would be 105 degrees, there would be five programmers and/or observers in the room, one manager, a coffee pot, one observer, no insulation in the walls or ceiling, and miscellaneous heat-producing equipment. The total heat generated came to 7.1 tons. To be on the safe side, they purchased a ten-ton unit.

They installed computer floors, which allowed them to put the cold air in the floor and have it blow up into the cabinets. Vents were also installed under each desk. This allowed the 52 degree air to blow up between the legs of the programmers, many of whom were females who wore skirts. In order to prevent frost bite, they stored the computer manuals on the vents.

On one project that was going into a Florida location, the programmers convinced three engineering directors that 10 tons of air-conditioning was slightly excessive and that a window unit from Sears would be more appropriate. The HVAC people compromised with a 7.5-ton unit. One day when the outside temperature was 96 degrees and the room had its full complement of people, we decided to test the compressor capacity by setting the thermostat as low as possible. When the temperature dropped to 57 degrees, the head of operations grabbed me by the neck and offered to put me out of my misery. I tried to explain to him that we had proven the air-conditioning people didn't know how to size the units. The next project that we put in Massachusetts had a 10-ton unit.

Back to the main subject————I now have this cute little six-pound box that has 20 meg of RAM and a hard disc in excess of 200 meg. How Wonderful!!!!!! It has little colored pictures of file folders, scissors, clip boards, envelopes, printers, etc., etc., ... It has menus with more menus. At least once a day a little box with a blue border pops up on the screen and says something like, TRANSMISSION RETRY VALUE IN PRINTERS CONNECT DIALOG BOX.......or.....OPTIONS—NETWORK SETTINGS—PRINT NET JOB DIRECT—OK? When you first encounter this kind of stuff, you look in the instruction books for some kind of guidance. NOTHING! The next step might be to call the vendor's 800 number. But which vendor? Some choices include:

- Microsoft. Hey, they did DOS, WINDOWS, WORD, or maybe
- Semantic. Hey, they did THE NORTON DESKTOP or maybe
- Hewlett Packard. They sent me the discs with all the fonts for the Laser printer.

Fortunately, my son lives in Mountain View, California—the heart of the computer world. Not only does he live in Mountain View, but he is an electrical engineer. Not only is he an electrical engineer, but he lives with two computer science people who work for big software companies near Mountain View, California. Not only do his two computer science roommates work for big software companies in Mountain View, California, but their girl friends work for other computer software companies. So, whenever one of these cryptic messages pops up on my screen, I call Mountain View and say, "what do I do?" And what do all these experts say?

- Don't look in the instruction book. You won't understand it.
- Don't call the software company. They keep you waiting on the phone.
- Shut off the computer. Maybe the problem will go away.

(2a) Why Doesn't the Desk top Manager Manage?

(3) How to Become Intelligent by Asking Stupid Questions.

Now that I am retired, companies hire me to teach their young, inexperienced engineers all the stuff that I learned before they paid me to leave. I start off my classes by suggesting that the students ask as many questions as possible. They are reminded that I will not be available after I get into the 39.6% tax bracket. They are also told that there are many really stupid industrial installations that don't work very well. If they don't understand why they were built the way they were, the young engineers will continue the practices and maybe cause some serious problems.

My favorite story is about my first boss. About 1960, most companies did not have full-time, experienced instrument engineers, so they used whoever was available. My boss was a mechanical engineer who knew what an instrument was. I was a chemical engineer who designed heating and air conditioning control systems during my freshman year, six-month co-op job. Together we shared an office and knew slightly more than nothing about instruments and controls. I read catalogs. He asked questions. He called up anyone who would talk to him and asked and asked and asked. He invited manufacturers, vendors, manufacturer's representatives, salespeople, technical advisors—anyone who said they knew anything about instruments—to visit him and explain what they were and what they did. The more questions he asked, the more embarrassed I became.

After about six months, a pattern evolved. He kept asking the same questions to the same people and also to different people. I was convinced that he was either stupid or just liked to ask questions. But, when you listened to the answers, you realized that some didn't make any sense and others were contradictory. He remembered all the questions, who answered them, and what they said every time. We both learned who knew what they were talking about and who was making up stories. Eventually we learned the instrument business, and I learned to keep asking questions until I found someone who knew what they were talking about and was willing to share the information.

MORAL: ASK A LOT OF STUPID QUESTIONS—OVER AND OVER AGAIN

(4) If a Computer Calls You During Dinner, Should You Be Rude?

(5) Why Won't the United States Government Let Me Use Their Telephones?

After working for many years for conventional (?) companies, I had the opportunity to become a "consultant." There is a very nice feeling about the concept of being asked for your opinion and getting paid for it — more later. However, consultants are treated in strange ways.

Story 1.39: I was asked to do some work for a small organization by the name of "THE UNITED STATES GOVERNMENT." You may have heard of them. The first meeting was held in one of those buildings that is used only to have meetings and classes. Thirty-five people walked into the room, found seats that were related to their expected participation, and waited for the facilitator to make announcements. The very first directive was that there were no phones available for general use. A long involved story tried to explain why phones were not available. Fortunately for me, the deli next door had a pay phone that was available during lunch time when they were open for business.

The following week some of us returned for another meeting in a different building that was owned and operated by another part of the United States Government. The first announcement was that there were no phones available for general use because they were all dedicated to computers, FAXs, or important people. Since I wasn't a computer, a FAX, or an important person and there was no deli nearby, I made my calls from the hotel at lunch time. Someone raised the possibility of moving the meeting to a building that had phones and more comfortable chairs. After about four hours of investigation, it turned out that the only other available building required that everyone have security clearance in order to leave the conference room. That meant that some of us couldn't use the phones, get a drink, or go to the restroom without a having a "trained and licensed" security person watching us.

We tried to figure out why there were no extra phones. Possible choices include the following:

- The US Government and Ma Bell don't like each other.
- The US Government is cheap.
- Ma Bell is cheap.
- The cellular phone companies have a deal going with the US Government to force people to carry portable phones.
- Phones are used to sneak information in and out of important meetings, so their use is discouraged.
- It's too complicated for any non-government engineer to understand.

(6) Why It's Better Not to Work and Make More Money.

(7) Going Back to School Is More Fun If You Are Smarter than the Professor.

(8) How To Be Productive Without Having Anything to Do.

(9) If You Give Them Money, They Will Ask for More.

(10) Are Smart Instruments Smarter than Engineers or Managers?

(11) If I'm Right, Are You Wrong? (or) Why Don't You Listen to Me, DUMMY?